# MEASURING
# DISTANCE

by Meg Gaertner

**Cody Koala**

An Imprint of Pop!

popbooksonline.com

abdobooks.com

Published by Pop!, a division of ABDO, PO Box 398166, Minneapolis, Minnesota 55439. Copyright © 2020 by POP, LLC. International copyrights reserved in all countries. No part of this book may be reproduced in any form without written permission from the publisher. Pop!™ is a trademark and logo of POP, LLC.

Printed in the United States of America, North Mankato, Minnesota

102019
012020

THIS BOOK CONTAINS
RECYCLED MATERIALS

Cover Photo: Shutterstock Images
Interior Photos: Shutterstock Images, 1, 8, 13 (top), 21 (measuring tape); iStockphoto, 5, 7, 11, 13 (bottom left), 13 (bottom right), 15, 16, 17, 19, 21 (basketball), 21 (football); Charlie Riedel/AP Images, 9

Editor: Meg Gaertner
Series Designer: Jake Slavik

Library of Congress Control Number: 2019942408

Publisher's Cataloging-in-Publication Data

Names: Gaertner, Meg, author.
Title: Measuring distance / by Meg Gaertner
Description: Minneapolis, Minnesota : Pop!, 2020 | Series: Let's measure | Includes online resources and index
Identifiers: ISBN 9781532165542 (lib. bdg.) | ISBN 9781532166860 (ebook)
Subjects: LCSH: Length measurement--Juvenile literature. | Distance perception--Juvenile literature. | Depth perception--Juvenile literature. | Measurement--Juvenile literature. | Mathematics--Juvenile literature.
Classification: DDC 530.813--dc23

WITHDRAWN

## Hello! My name is

# Cody Koala

Pop open this book and you'll find QR codes like this one, loaded with information, so you can learn even more!

Scan this code* and others like it while you read, or visit the website below to make this book pop.

**popbooksonline.com/measuring-distance**

*Scanning QR codes requires a web-enabled smart device with a QR code reader app and a camera.

# Table of Contents

# What Is Distance?

Distance is the **length** of a straight line between two objects. People measure distance to tell how far apart two objects are.

distance

Watch a video here!

# Units of Measurement

People use several units to measure distance. A basic unit of measurement is the **foot**.

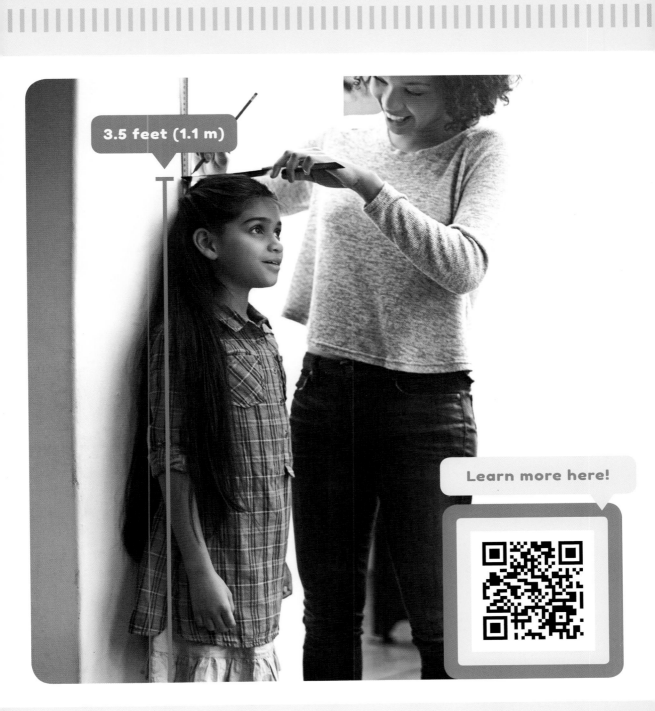

3.5 feet (1.1 m)

Learn more here!

one inch

For very short distances,
people measure in **inches**.
One foot is 12 inches
(30.5 cm).

one yard

For longer distances, people use **yards**. One yard is 3 feet (0.9 m).

A football field is 100 yards (91 m) long.

For very long distances, people measure in **miles**. Very long distances include the distances between two cities.

# Chapter 3

# Tools for Measuring

People use different tools
to measure distance. All
of these tools have lines
or markings on them.
The lines show the units
of measurement.

Learn more here!

**Rulers** are useful for measuring short distances. Most rulers are 1 **foot** (30.5 cm) long. The long lines on a ruler mark each **inch**.

ruler

yardstick

People can also use yardsticks or measuring tapes. A yardstick is exactly 1 **yard** (0.9 m) long.

measuring tape

A measuring tape is a long, thin ruler than can bend. Both yardsticks and measuring tapes have lines marking each inch.

# Chapter 4

# Measure It!

To measure the distance between two objects, first choose a measuring tool.

Complete an activity here!

Place one end of the tool at one of the objects. Bring the other end of the tool to the other object. Read the distance by finding the closest marking on the tool.

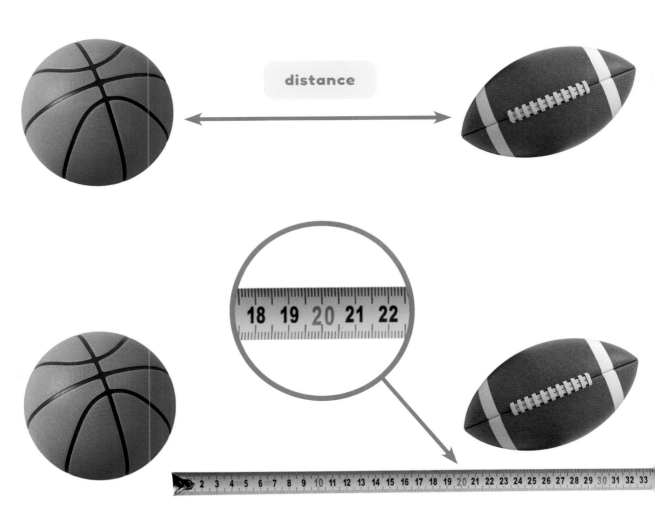

distance

# Making Connections

## Text-to-Self

Have you ever seen someone measure distance? What tool did that person use?

## Text-to-Text

Have you read other books about measuring? What did you learn?

## Text-to-World

Eve wants to measure the distance between the door and the window of her bedroom. Should she measure in inches, feet, or miles? Why?

# Glossary

**foot** – the basic unit of length in the US standard system of measurement.

**inch** – a unit of length that is smaller than a foot. One foot is 12 inches (30.5 cm).

**length** – a measurement from one end of something to the other.

**mile** – a unit of length that equals 5,280 feet (1.6 km).

**ruler** – a straight piece of hard material that is used for measuring length.

**yard** – a unit of length that equals 3 feet (0.9 m).

# Index

## Online Resources

# popbooksonline.com

Thanks for reading this Cody Koala book!

Scan this code* and others like it in this book, or visit the website below to make this book pop!

**popbooksonline.com/measuring-distance**

*Scanning QR codes requires a web-enabled smart device with a QR code reader app and a camera.